Photography

I0503414

Guide to Taking Stunning, Beautiful DSLR Photography and Smart Phones

By: Jonathan Cooper

FREE BONUS: "Click The Link Below To Receive Your Bonus

https://publishfs.leadpages.co/pangea-health/

Table of Contents

Introduction

Photography envelops science, art, application and practice. Maybe it is precisely this breadth of human knowledge and skills what makes photography so popular and widespread. Everyone should become a photographer. By becoming one, you could capture a moment that would otherwise be gone forever, you could start to see the beauty everywhere around you, you could document your family and the people you care about, you could express yourself and become creative, you could make other people think and communicate to them your impressions, and you could simply enjoy the wonderful moments of your life. There are many reasons to become a photographer, but would you like to become a good one?

There is a huge difference between someone who is taking photos and the one who is making them. Your camera is just an instrument, but you are its master. You are the artist. In this book you will find some great insights on the photography and the most useful guidelines in taking beautiful and stunning images. It takes time to learn well all these important rules and to put into practice all that you have learned. Henri Cartier-Bresson once said that your first 10,000 photographs are your worst. But without these rules and guidelines you will not be able to advance. Without consulting them, you would be like someone walking into the forest not knowing where the sides of the world are. But there is even much more to it. A great photographer doesn't only master his settings. He is also capable of expressing his deep feelings about what he is photographing.

Photography has become one of the most present arts and skills in our world. Think only about all the images that connect millions of people through social media posts. This book was written to help you improve your photography skills and enable you to take a beautiful pictures you will be proud of. There is a secret in every great picture. But there are also secrets in the ways of taking the great pictures. So get to know them!

Chapter 1: Understanding Your Camera

When you decide to make a step forward and take your camera off 'Auto' mode, you will find yourself in front of unknown settings. It is necessary to consider all of them right away. The exploring and the knowing of the effects of each setting will soon allow you to be in control of your camera, understanding how to achieve the photographic results.

Exposure

To be able to take control of your camera and to take good photos, you will need to learn how exposure works. Exposure is combined of three elements, usually called the exposure triangle:

- Aperture – the size of the opening of the lens when a picture is taken.
- Shutter speed – the amount of time that the shutter is open.
- ISO – the measure of digital camera sensor's sensitivity to light.

Each of them relates to the light and on the way the light enters and interacts with the camera. A change in one of the elements impacts all the others. So you will need to keep all of them in mind, without ever isolating just one of the elements. Moreover, each of the three aspects of the triangle will not affect only the exposure, but also cause alterations in the depth of field, motion blur, and digital noise.

1. Aperture

Aperture is the property of camera lenses, similar to the eye pupil that opens or closes to let light pass. The aperture refers to the diameter of the hole inside the lens. The alteration of the size of this hole allows more or less light into the camera. The wider the aperture, the more light is allowed in and vice versa.

In photography, aperture is expressed in f-numbers, known as "f-stops". This is the way of describing the size of the aperture, or how open or closed the aperture is. A smaller f-stop means a larger aperture, while a larger f-stop means a smaller aperture. The scale is ranked as follows:

f/1.4, f/2, f/2.8, f/4, f/5.6, f/8, f/11, f/16, f/22

The aperture size shrinks as you move from f/1.4 to f/22, while as the aperture widens, the f/number gets lower and more light is allowed into the camera. This will do well for the low light, but it's going to make the depth of field very shallow, which is not ideal for landscapes.

This change in aperture directly effects the depth, which is the area of the image that appears sharp. A large f-number brings all foreground and background objects in focus, while a small f-number isolates the foreground from the background by making the foreground objects sharp and blurring the background.

2. Shutter speed

The light that has passed through the lens aperture hits the camera shutter. The shutter is the curtain that is in front of the sensor. It will stay closed until you are ready to fire the camera. It opens when the photograph is ready to be taken, and fully exposes the sensor to the light. After the sensor has collected the light, the shutter closes immediately.

Shutter speed or exposure time is the length of time a camera shutter stays opened to expose light into the camera sensor. The fast shutter speed can completely freeze the action, while the slow shutter speed can make moving objects appear blurred, an effect called "motion blur". Slow shutter speeds are also used to photograph objects at night or in dim environments with a tripod, or to create a sense of motion in landscape photographs. Shutter speed thus allows you to create dramatic effects: high shutter speeds freezes action, while low shutter speeds creates an effect of motion.

Shutter speeds are measured in fractions of a second or in a seconds. Different shutter speeds correspond to different situations: from really fast (1/8000th of a second and faster) for sports photography, to really slow (30 seconds) for low-light or night photography, or to capture the movement. Ordinarily you will want a very small fraction of a second to prevent motion blur and to freeze action. This is usually above 1/500th of a second for general photography and above 1/1000th of a second for sports photography. Slow shutter speed is the slowest one that can be handled without camera shake. Long shutter speeds are normally above 1 second, and you will have to use a tripod to get sharp images.

To automatically set shutter speed, you can use the 'Auto' mode, while in 'Aperture Priority' mode, you will have to set the lens aperture, while the camera will automatically set the shutter speed. For the manual setting of the shutter speed, you can choose between 'Shutter Priority' mode, where you set the shutter speed and the camera automatically selects the aperture, or the 'Manual' mode, where everything is set manually. It is usually the best option to use the 'Aperture Priority' mode.

3. ISO

The light that has passed through the aperture and has been filtered by the shutter speed, now reaches the sensor. Image sensor or simply sensor is the most important and most expensive part of a camera, responsible for the gathering of light and transforming it into an image. There you decide upon the ISO or the level of sensitivity of the camera to available light. The lower the ISO number, the less sensitive it is to the light. A higher ISO number will raise the sensitivity of your camera, allowing you to capture pictures when in low-light environments avoiding the use of a flash. As you turn the ISO number up, you increase the exposure, but the image quality will decrease. You will introduce more digital noise or 'grain'. You will have to decide is this grain acceptable for your image.

The sensitivity of the ISO is measured numerically from low sensitivity (ISO 100) to high sensitivity (ISO 6400) and depending on the camera, the range could be more. The ISO sequence is: 100, 200, 400, 800, 1600, 3200, 6400 and etc. You need to understand, that at each step between the numbers, the sensitivity of the sensor is doubled.

Practically speaking, you will usually want to keep the ISO as low as possible (ISO 100 or 200), especially when there is plenty of light. This will allow you to retain the most detail and to have the highest image quality. If it is cloudy, you could select an ISO between 400 and 800, and if you shoot indoors, consider an ISO of around 1600 or above, to be able to capture the moment without introducing blur to the image. You might also want to increase ISO when getting ultra-fast shots.

Shooting Modes

Photographers are able to control the parameters of the exposure by using digital camera modes. Certain modes fully automate the camera exposure, while the others let you manually control some or all of the parameters. Most digital cameras have different types of modes for various situations, allowing both automatic and manual exposure control.

The four main types of camera modes that can be found in most digital cameras today are: (P) Program, (Tv) or (S) Shutter Priority, (A) or (Av) Aperture Priority, and (M) Manual.

1. Program mode (P)

In program mode, the camera adjusts both the aperture and the shutter speed, based on the amount of light that passes through the lens. It automatically produces what it judges to be the best exposure for the object you are photographing. This mode allows you also to change to a different combination of aperture and shutter speed. It is mostly used for the occasions when you just need to take a quick shot.

2. Shutter priority (Tv) or (S)

Shutter priority is a semi-automatic shooting mode. You will need to manually adjust the shutter speed of the camera and it will determine the appropriate aperture required to give the correct exposure. If there is too much light, the camera will automatically increase the lens aperture to a higher number, decreasing the amount of light that passes through the lens. If you are in a low-light environment, the camera will decrease the aperture, so that more light can pass through the lens. This mode is used for the freezing of the motion or for the intentional blurring.

3. Aperture priority (Av) or (A)

Aperture priority is another semi-automatic mode. You need to manually adjust the aperture of the lens, and the camera will automatically pick the right shutter speed to give the proper exposure. If there is too much light, the camera will automatically increase the shutter speed, while if there is not enough light, the camera will decrease the shutter speed. This mode allows you to have full control over subject isolation and to play with the depth of field. At the same time, there is no risk of overexposure or underexposure.

4. Manual mode (M)

In manual mode you are given full manual control to determine the exposure. You can manually set both the aperture and the shutter speed to any value you want. This mode is generally used in extreme lighting situations, when the camera is not arriving to figure out the correct exposure.

Types of Lenses

The five essential types of lenses are: standard zoom, zoom, wide angle, close-up and macro prime.

1. **Standard zoom lens** is the most common type, with the general focal length of 18mm-55mm, useful for amateurs and beginner photographers. These types of lenses have a variety of options for composition.

2. **Zoom or super zoom lens** has a much longer focal length and there are many different choices. That's why they are great for long-range photography like wildlife, sports or music.

3. **Wide angle lenses** are very popular for landscape and nature photographs. Fish Eye lens is a special interest lens that distorts a wide-angle view into a circular image.

4. **Close-up or macro lenses** are the specialist lenses made for close focusing. They are good for the photos of insects, snowflakes or details.

5. **Prime lenses** have fixed focal lengths. This means that you cannot adjust the zoom of the lens, but you must physically move closer or further away from the subject. They are generally considered to give the photos of the highest quality and sharpness, but are also considerably costly.

Lenses are usually tested according to the sharpness, distortion, vignetting and chromatic aberration. For you it is also important the aperture value, which is related to the depth of field.

Other Basics of Photography

1. White balance

The white balance can significantly change the color tone and the overall warmth of your photographs, making them appear blue or orange, warm or cold. This depends on the color temperature of the light sources. It is recommended to avoid auto white balance, since it can leave you with wrong colors. The best is to set it manually, making your photos look more accurate. With jpeg files the white balance can't be adjusted in post-processing, so you need to do it before taking an image. By looking at the kind of the day or emplacement, you can easily determine and select the appropriate color balance. 'Daylight' is used on the bright sunlight. 'Cloudy' ads warm tones on a cloudy days. 'Shade' warms up the images taken in the shade. 'Tungsten' cools down the yellow tones when shooting indoors and under incandescent light bulbs or street lights. 'Fluorescent' makes up for green or blue tones of indoors fluorescent lights, and 'Flash' ads a cool blue to the images.

2. Metering and metering modes

Metering means the calculation of an average exposure. Your camera will determine correct shutter speed and aperture depending on the amount of light and the sensor's sensitivity. The most common metering modes that you can choose from are: matrix or evaluative metering, center-weighted metering, and spot metering. In matrix or evaluative metering the camera will analyze the light and dark tones of the entire frame by dividing it into different zones, and determine the exposure so that all of the tones average to 18% grey. Center-weighted metering evaluates the light in the center of the frame that can total up to approximately 80% of the scene, ignoring the corners. Spot metering evaluates the light a very small area of the scene that totals approximately 5% of the viewfinder area, and ignores everything else.

3. Underexposure and overexposure

An underexposed image is the photograph one might consider to be too dark, while the overexposed image one might consider to be too bright. To increase or decrease the camera default meter reading, there is a small +/- button near the shutter. For the scenes with primarily dark tones you can use the negative exposure compensation, and for the scenes with primarily bright tones you can make your camera know that the scene should be brighter.

4. Histograms

A histogram shows a graphical representation of the tonal values of you photograph. This amount of tones of various brightness levels ranks from black (0% brightness) to white (100% brightness). This doesn't mean that the histogram is to be used for evaluation of the exposure. You can use it to avoid loss of highlight or shadow details at specific exposure settings.

5. Focusing

Focusing allows you to choose what you are focusing on, ensuring that the subject you want to capture is in focus of your photo. You can do it by setting your camera to the basic focusing modes and focus point selection, or you can explore the more advanced modes that your camera offers.

DSLRs offer a variety of autofocus modes. The two most important ones are AF-S and AF-C autofocus modes. AF-S (autofocus-single) is used for portraits, landscapes, architecture and other stationary subjects, while AF-C (autofocus-continuous) for moving ones like wildlife or sports. In both cases, the camera will initially acquire focus on the particular focus point when you half-press the shutter. To change the focus in AF-S mode, you need to release the button, recompose and then half-press again. In AF-C, the camera will utilize the surrounding focus points to track subject movement and keep focus on your subject until the photograph is taken. The focus modes rely on focus points, meaning the little squares or dots that you see when you look through your viewfinder. The active focus point is highlighted when you half-press the shutter. To change the selected single focus point you can use the directional buttons.

6. File size or types

You will have the option to choose which size and file type of the images you would like to record. The file size can be large, fine or super fine, based on the mega pixels. The file type could be raw or jpeg. A raw file is uncompressed, allowing you a huge flexibility during post-processing. A jpeg is a compressed file type, which is automatically processed by the camera. This file type is more recommended for the beginners.

7. Crop factor

The size of the sensor directly influences the field of view and the resulting image. A sensor that is smaller than professional SLR cameras will essentially create a narrower viewing angle and crop the image.

8. Focal length

The choice of focal length can affect the zoom and the perspective. It affects not only what you see but also how you see it, because different focal lengths have different fields of view. A longer focal length is taken into consideration for details that can stand on their own. A shorter focal length is for bigger picture scenes like buildings or landscapes. A normal focal length would be 35mm or 50mm.

9. Polarizing filters

Polarizing filters make light enter the lens only from a certain direction, thus removing the glare and reflections from objects like water and glass, as well as the haze from the sky. This cannot be replicated in post-production.

Chapter 2: How to Take Sharp Photos

One of the most disappointing and frustrating moments when taking a picture is when it comes our soft, blurry or out of focus. The following techniques will help you take sharp images. The main reasons for getting blurry images are slow shutter speed, high ISO number, poor focusing, moving subject, bad lens, and improper hand-holding techniques.

1. For the beginning, set your camera to the ISO base value (ISO 100 or 200), which results in the highest quality images with maximum sharpness and less noise. You could even set your camera to automatically change the sensitivity of the sensor according to the available light. Do it by turning your Auto-ISO feature to 'On' by using the following settings: ISO auto control 'On', max sensitivity: 1600, and minimum shutter speed: 1/100.

2. In the case that your zoom lens goes beyond 100mm, remember that the shutter speed should be at least equivalent to the effective focal length set on the lens.

3. Pay attention to the aperture. Shooting between f/5 and f/10 generally produces the best results, regardless of lens. Smaller apertures can soften the image, while large apertures reduce your depth of field, leaving some parts less in focus than others.

4. It could be practical for you to adjust your camera to the aperture-priority mode and decrease the aperture to the minimum possible value.

5. If you set your camera to matrix or evaluative metering, the shutter speed will be measured upon the analysis of the entire frame.

6. When you half-press the shutter while you point your camera to the subject you want to photograph in the aperture priority mode, you will be able to see the shutter speed on the bottom of your viewfinder. It should be 1/100 or more. If it is below that value, increase it by allowing more light to come in.

7. Try to increase the shutter speed number to ISO 800 or even higher. The lower shutter speed can make your photos look blurry while holding your camera in your hands.

8. Improve your hand-holding technique. Relax and take a steady position.

9. Make sure that your anti-shake or anti-vibration reduction technology is 'On'.

10. Learn how to focus correctly. Pay attention to the difference between a blur caused by camera shake and the focusing issues. A slow shutter speed or camera shake results in blurriness of the whole image, while the problem in focusing makes the background of your image sharp, leaving the subject out of focus. This can be caused by the lack of light or misplaced focus point.

11. Always focus on the closest eye of the person or animal you are photographing.

12. To avoid motion blur while taking photos with slow shutter speed, make your subject stand still.

13. A good quality lens, like fast prime lens, can have a major impact on the sharpness of your images. The more expensive lenses are generally sharper than cheap ones.

14. Remove lens filters when not needed to improve the clarity of your images.

15. Pay attention to the aperture to see if it makes a difference in sharpness. The smaller aperture requires the longer shutter speed, making it more difficult to keep sharp the moving subjects.

16. Keep your lenses and all the equipment clean.

17. Use a tripod or other sturdy surface for the situations of low light to reduce camera shake.

18. Use a remote cable release or remote control to avoid shaking.

19. Block wind on the windy weather.

20. You can also sharpen your pictures in post-processing. You could manually blend your exposures, use sharpening tools, or sharpen your image and remove noise selectively.

Chapter 3: Composition

Composition means the placement of objects and elements in a work of art, which directly makes impact on its quality. Thanks to the composition, a photographer directs the viewer's eye towards the elements he judges to be the most important ones of his work. That's why it is essential to choose well your composition. Composition elements are shapes, tones, shadows, highlights, colors. Composition is also affected by the position of your camera, angle of shooting, focal length, aperture, etc. There are certain rules for the composition in photography. You are not obliged to follow every one of them, but they are the great guides in helping you to take better photos.

The Rule of Thirds

Rule of thirds comes from the well-known golden ratio, which is considered to define perfect proportion. Basically, you should divide your camera's frame into three equal parts or into nine equal rectangles by four intersecting lines (two vertically, and two horizontally) and place the elements of importance at the intersection points of these line. The reason is that the human eye tends to be more interested in the subjects that are placed at those divisions.

The Rule of Odds

People prefer the images that contain an odd number of elements rather than an even number. It is because human eye spontaneously goes towards the center of a group, no matter what have you placed there. So the best thing for you as a photographer, is to make your viewer look at the subject of your choice.

Visual Weight

Visual weight is about what we are drawn to when we look at a photograph. The understanding of the visual weight is essential for the understanding of how people look at photograph. This will determine the way you position certain elements in a frame in order to direct the viewers' attention where you want them to look.

Lines

1. **Aligned Lines and Straightened Horizon**

   People naturally like straight things or a visual balance. When you take a pictura, look for anything that might serve as your visual guide. Identify the horizon or the lines by which the scene needs to be adjusted and aligned. Then you can start framing your subject. Sometimes it can be difficult to find a visual guide, but try not to leave the aligning for the post-processing, because you will not be able to do it well enough.

2. **Leading Lines**

Creation and use of leading lines is one of the most important elements of the composition. The human eye is drawn into image along lines, and taken wherever you want to lead it. The lines can be straight, curved, diagonal, etc. without clear lines or something else that will lead the viewer's eye, he might wonder around the image without landing on a particular spot. Leading line gives a sense of visual flow, balance and depth. Many different elements and compositional factors can become leading lines, entering from all the corners of the image or forming an interesting pattern.

Patterns and Texture

We find patterns everywhere around us, whether they are man-made or the natural ones. Patterns enrich the image with the harmony and rhythm, creating the sense of order and peace. If you break the rhythm of the pattern on your image, the eye will fall upon the specific point, returning afterwards to the harmonious rhythm of the pattern.

Texture can help you create dimension in your image, making it more visually interesting because of appearance of depth.

Colors

Colors are the most powerful means for creating and communicating the particular feeling or mood. You can use the differences between warm and cold colors, or you can play with the sudden contrasts to dramatically change the viewer's perception and make him feel the way you want him to feel when looking at your image.

Symmetry

A symmetrical image looks the same on both sides. You can use a symmetry in two different ways. The first one is to find the symmetrical patterns with defined lines, and the second one is to find the symmetrical patterns in unexpected, mostly natural places. The beauty of the chosen subject is the important element in taking appealing symmetrical images.

Directing the Eye

To direct or to lead the eye of the viewer into the image and through the scene is an important part of the composition, making it more interesting and stronger. This can be done in various ways. You could use the physical features horizon line, roads, coastlines, buildings, bridges, furniture, stairs, patterned floors or even people. It could be done by the very subject or the location, by the emplacement of the focal point, by using the light and the shadow.

Balance

While taking his image, the photographer decides if the composition is going to be balanced or imbalanced. His choice will affect the way we feel when we look at it: relaxed, calm or uneasy. A photograph is considered perfectly balanced when left and right sides draw the eye equally. The top and bottom halves are of little importance when it comes to balance. This can be obtained through the symmetry, but even the asymmetrical photograph can be balanced. The reason is that an eye can be attracted by diverse elements or aspects, like areas of contrast, bright spots, focused items, warm colors, large items, the eyes and the direction of the gaze.

Often it is necessary for a photograph to be imbalanced, especially if you want to show the movement and the tension of the scene. The imbalance can improve the quality of your image by making the composition more intentional and stronger.

The good thing about balance is that you can highlight certain aspects of the frame in the editing stages, after the pictures are taken.

Types of Composition

There are two main composition types and several smaller branches. The two main ones are opened and closed composition. In the closed composition, all the elements are arranged inside the frame, not drawing the viewer's eye away or making it jump from one object to another. The main object is clearly highlighted, while the other elements direct viewer's eye towards that object, usually placed near the center of the image. Such compositions give static and stable images that feel complete and calm.

An open composition, on the contrary, transmits the sense of movement. The dynamism of such images is obtained through different means, such as placement of objects, plenty of lines, colors and elements, making the viewer's eye go from one object to another. The elements of the image are not constrained, but they go towards the edges and seem to go beyond.

Sometimes it can be difficult to discern the type of composition of certain photographs, having the characteristics of both types. Still, characteristics of one type will be a bit more present. The works of art simply go beyond the predefined rules and definitions.

Central Composition

People spontaneously place the main object of the image in the very center of their frame. But you will often hear an advice not to put your object in the middle, but rather close to the left or right third of the frame. It is true that such a composition can make your photograph look more creative. Nevertheless, the central composition offers you great and interesting possibilities, if you learn how and when to use it properly. It is often calm and static, and it feels like introducing the object to the viewer and leading the viewer's gaze where you want it to go.

Negative Space

Negative space is the space that surrounds the element or the object of interest in your image, making it stand out and attract the viewer's attention. You can create the negative space with all sorts of elements, like emptiness, buildings, crowds, or with the help of light. The beauty of such a composition is that it makes you create and communicate a particular emotion or mood.

Simplification

Simple images are generally more appealing than complicated ones. That's why it is good to get rid of distracting, unnecessary elements in your image that add nothing to your composition. There are several ways to do it: by zooming in on your subject, recomposing without the element in the frame, using a wider aperture, or cropping the image in post-processing.

Viewpoint and Perspective

Viewpoint and perspective can dramatically change the viewer's perception of an object and the mood of a photograph. For example, if in shooting a portrait you place yourself above the person, you will make her appear insignificant, of lesser value than the viewer. But if you shoot that person from below, she will appear dominant. In changing the perspective, you can play with the appearance of the size of the object, making it look higher or smaller.

Background

Beginners often forget to pay attention to the background. There are backgrounds that add something to your composition, and there are those who don't. Ask yourself whether they will contribute to your final image, and include them or exclude according to your answer.

Depth

The choice of depth depends on the type of image you are trying to capture, whether it is a landscape or a portrait. You can add dimension in several ways: by including or excluding the background, by including something in the foreground, or by overlapping certain elements.

Chapter 4: Improve Your Portrait, Landscape and Nature Photography

Portrait Photography

Portrait photography allow us to preserve the unique life moments and to capture true personalities. This takes more than just high quality photography gear, and even more than simple understanding of the photography principles. Portraits are about relations. While taking a portrait, you are engaging yourself in a personal relation. The result of your work will depend on the way people feel and behave in front of your camera.

1. Build a relation

To make your subject feel comfortable, do your best to connect. Draw a smile and start a conversation to become familiar. Before you start shooting, don't just keep silent but offer them direction. Explain to them what you want from them and how you want them to pose. If you are photographing the unknown people on the street, ask them for the permission. Some of them will not be in favor of being randomly photographed. Politely let them pass without photographing them. Be particularly careful when asking the permission for photographing the children. Parents can be very protective of their kids. At the end, don't forget to compliment and thank the person who has agreed to be your subject. Also show them the photos you've taken. In this way you will build the confidence.

2. Pay attention to the lighting and environment

Consider the time of day and the position of the sun. Early morning and late afternoon are the best times for a natural-light portraits. On cloudy days you will be getting portraits with softer shadows. The best would be to make your subject face the sun, or to have his defined features lit at an oblique angle. Also be aware of the background. Look beyond your subject to check the composition issues so as to avoid unwanted tensions and distractions.

3. Position your camera

In shooting a portrait, you will want to capture a pleasing one. So you will need to make the best of your subject's features, no matter how pronounced or heavy they could be. It is therefore crucial to well position your camera and to choose the lens and the focal length that will lead you to successful results. Keep in mind that elements will appear larger the closer they are to the camera. This effect will be amplified with the use of wide angle lenses.

4. The importance of the eyes

The eyes are the most important element of the portrait. They reveal the personality. That's why you should always emphasize the eyes. If the eyes aren't sharp, something huge will be lacking. Photographers often choose to place the focusing point right on their subject's eye, thus keeping it always sharp regardless of how shallow the depth of field could be.

5. Move around and keep the eye level

To explore well your subject, move around, both with your feet and with your lens. Zooming in and out will allow you to capture a lot of environment or to obtain a more intimate portrait. Also keep your camera at the eye level, especially when shooting children.

6. Check the camera settings

Watch your exposure, meaning the aperture, shutter speed and ISO settings. Choose the best lenses for portraits, like wide-angle lens. And pay special attention to the white balance. The altered tonality and appearance of your subject's skin can give you undesired results.

7. Create interesting compositions

Sometimes it is good to follow the habitual compositions of including all of your subject or the top half. But very often it is better to search for an inspiration and become creative with your compositions. You can learn to master some great techniques, like positioning the subject to one side of the frame, with enough space in front to look into.

8. Play around with posing

The poses and looks dramatically affect the shooting results. Even the slightest changes in facial expressions can communicate very different feelings. Try to capture a range of expressions and let your subject look in different directions. Don't hesitate to play around and see what works.

Landscape Photography

To shoot the great landscape photos, first you will need to find the beautiful locations. Then you will have to figure out at what time of day and in what season they will appear in all of their splendor. The next step will be to improve you technical and artistic skills to make the most of the scene. This will come mostly with the practice, but here are some guides to help you start and to show you how to progress.

1. Learn in advance about your landscape

Sometimes you can get lucky and unexpectedly discover the great landscape, but most of the time you won't. That's why it is good to inform yourself about the location, weather or sun position before you go. A scene can change dramatically depending upon the weather at any given moment. You can use online map services with satellite view to gather basic insights, and check out the online photos to learn what to expect from a location and to choose the precise spot. You should also inform yourself on other practical aspects of your journey, like access, paths, and parking spots. You can also visit online sites or apps to check when and where the sun will rise and set, especially if you have chosen to shoot photos on the mountain.

2. Be flexible

Regardless the detailed planning, there is always possibility of something unexpected to happen. The nature is simply too unpredictable. There could be the unexpected changes in the weather, but also in the landscape itself, like changes in the flora, water levels or courses. You could even encounter more important changes after dramatic events. That's why you should always be ready for the change of plans. The good thing is that these unexpected changes can bring forth great inspirations and new ideas, producing even better results than what you have originally thought.

3. Check your camera settings and your lenses

With landscape photography, you will normally want to get as much of your scene in focus as possible. To do this, choose a **small aperture setting (a large number), which will result in the great depth of field in your shots.** This lessening of light hitting your image sensor you will need to compensate either by increasing your ISO or lengthening your shutter speed (or both).

Chose a focal point for your landscape images, be it a huge tree, rock, building, interesting structure, etc. Think also where you want to place the chosen focal point.

The wide-angle lens is the most used one for the landscape shooting, but if you want to capture more creative images, there is a possibility to use also zoom and telephoto lenses.

4. Work with the weather

Beginners may sometimes think that the day has to be sunny in order to go out with their camera. But the truth is that storms, wind, thick clouds, rainbows, or sunrises can offer you great opportunities to create dramatic images. The light, colors and textures before and after a storm offer some of the most impressive conditions for landscape photography. Just be careful and make sure that you and your gear are safe and well protected.

5. Explore your location

Landscapes usually offer many viewpoints that you can use. Take your tame and explore the environment. Experiment with different viewpoints, and maybe you will find something very unique.

6. Consider the foregrounds and the sky

Think carefully about the foreground of your shots. They are of real interest because they create a sense of depth on the images and draw in the viewer's gaze. The sky is also an important element. If it is beautiful and dramatic, make it to dominate your image. On the contrary, if the sky is plain and boring, place it in the upper third of your picture.

7. Use lines

Lines help you lead the eye of the people who view your image. Landscapes can provide you with interesting lines, which will give your images depth and create patterns.

8. Capture movement

Landscapes offer calm, serene and passive environments, but also those filled with drama, power and strong feelings. In order to capture the movement, set a long shutter speed and compensate it with the small aperture. This way you will regulate the light hitting your sensor. You could also use some sort of a filter.

9. Think about horizons

There are two things to take into consideration about the horizon before you take your landscape shot. The first one is if the horizon is straight. It is easier to straighten it right on spot than in post-processing. The second one is where to place it in your composition. The rule of thirds is usually applied here, whether on the top third line or the bottom one – depending on the importance of the sky for your image.

10. Use a tripod

As a result of the longer shutter speed and a small aperture for taking landscape images, you will have to ensure that your camera is completely still during the exposure. Wireless shutter release mechanisms or a cable are options when trying to obtain a perfect level of stillness for your camera.

Nature Photography

If you love nature and wildlife, make them your favorite photographic subject. Photographing nature can be challenging. But the effort you will put behind your photos will be compensated by the memories of travel, the senses evoked in particular places and at particular moments, and by the simple joy of life.

1. Understand light

Light is what makes photography possible. You need to learn to see the light and to use it to your best advantage. Light can transform your subject or influence the viewer's perception of it. You could find yourself in the situations where the light isn't ideal, but even this can add a mood to your image.

Pay attention to the direction of the light, on the way it falls on your subject. Is it from the back, side or front? Direct front light is the one that is behind you, reaching over your shoulders. It illuminates all parts equally and is often warmer. It is a general rule to put the sun behind you. Sidelight creates shadows and can add a dramatic ton to your images. If you point your camera toward the sun, you will get the overly bright areas and silhouettes of everything else.

Light also has its character. It can be hard or soft, depending on its source and on the weather. Hard light creates strong contrasts with dark and clear shadows. The low contrasts of the soft light are created by clouds, fog, haze, dust, or the atmosphere itself at the sunrise and sunset. You will have to decide what kind of light you want for your image.

Light has wide color range, from warm sunrise and sunset reds to cold midday blues. Pull them into your picture to create impact. The heavy, clouded sky will also emphasize the cold colors. Our eyes are drawn to vivid colors. Warm colors advance towards viewer, while cold colors recede from the viewer. That's why they can be used to bring the dimension into your image, like when having warm highlights, with cooler shadow areas. Color can also provide a powerful focal point.

Light continuously changes, making your subject appear very differently at each moment. It changes according to the season, time of the day, and even as quickly as moment to moment. These variations will also produce diverse emotional responses in the viewer.

Photographers usually advice to keep to the hours of golden light. This means being at the spot before sunrise and making the most of the last hours of sunlight. The midday sun is generally harsh, with exception on the overcast days when you are free to shoot all day long.

2. Know your subject

Nature and wildlife photography is often about capturing interesting poses or behavior. Being able to know your subject and be able to predict their behavior on some level, will be beneficial to the shot. Every animal species has certain patterns of behavior you could learn to know beforehand by spending your time with them. You need to learn to be patient, to just sit and watch. In this way you will become ready and prepared to capture the unique natural moments.

3. Think about the composition

Pause and ask yourself what you are trying to show. Is it a mountain range, a river flow, or a plant detail? Your choice of what you include and exclude will determine what is important. Try to imagine the picture before you take it. Your composition will depend on where you stand, where you point your camera, and whether you zoom in or out. Often it is the best to use the longest or the biggest lens possible, but you can also experiment with a wide-angle lens to be able to capture and show the environment. No need to say that for shooting wider you will have to come really close to your subject. Also, try to put yourself at an eye-level perspective. This will bring the viewers of your image right into the scene and place them in your subject's perspective.

4. Get familiar with the settings

You need to get to know your gear. The awesome nature and wildlife moments on average last only few seconds, and it would be pity to miss the precious moment or to blow the images. Learn how to make most of the changes that will be needed to your exposure and focus settings so that you can keep your eye viewing the shot through the viewfinder. Get to know about the shutter speed and ISO settings, know the added margins and focus points.

The easiest way to ruin your photograph is by making it to be blurry. Learn well how to obrain the sharp image. For far-off subjects like a horizon or distant mountain, it is not a problem. But be attentive when you are shooting something that is close to you, like a beautiful wildflower. By pressing the button halfway down, the camera will focus. Even the most of smartphones have this possibility to focus on a chosen spot by touching the screen before taking the picture.

The following tips will offer you additional help in learning what to take into consideration when photographing the nature or wildlife:

- Get up early and be on the chosen spot during the moments of the early morning light. This will also make you avoid the crowds that could gather later on.

- Learn to appreciate the diffused light of the overcast days.

- Know the rules and follow them, but also feel free to break them when you can create something unique and personal.

- Don't always look for the exceptional content. There are great photograph opportunities everywhere around you.

- Be attentive on your movements while photographing nature. Don't trample the plants and never uproot or cut the wildflowers.

- Use textures revealed by the early or late sun to communicate your feelings about the scene.

- When photographing wildlife, opt for a natural background without artificial objects.

- Use a tripod for lenses longer than 300mm, unless you want to get the serendipitous effects.

- Give a voice to your image. Think of adjectives to describe it and add them to the image.

- Be careful about the background while shooting the wild animals. They usually blend into the landscape. You can also blur out background distractions by using a shallow depth of field for close-ups.

- Pay attention to the exposition while taking pictures of the snow. The bright white can fool meters.

- When photographing details, look for the most interesting compositions. You can try different angles: above, below, from the side.
- Take photos of the wildflowers at the sunset. The softness and the warmth of the light will make them glow.
- Arm yourself with patience when looking for the dramatic and rare natural scenes like lightning. Nature is very unpredictable.
- Simplify things. The most of strong pictures have very few elements in them.
- Love to experiment. Explore the nature and all the possibilities it offers you for taking great pictures.
- Enjoy to be in the nature. Let its beauty be imprinted not only on your images, but also on you.

Chapter 5: Mistakes Often Made By DSLR Camera Users

People make mistakes at all the levels of the photographing process. Some of these mistakes are caused by simple human distractions, like forgetting their gear at home or not recharging the batteries. But there are also very common mistakes of technical nature, caused by the lack of knowledge, skill or practice.

1. Unrealistic or too strong colors

The poor color choice is usually due to the lack of knowing any better or the lack of experience. It can also result from the attempt to make your photos look like paintings, but instead making them look fake. It would be very useful to get a monitor that is color calibrated. Also take into consideration that the printed image has stronger colors than the ones you see on a monitor. Simply try to avoid a large amount of color saturation and make the best possible attempt to have your photos seem as balanced and as close to natural as you possibly can.

2. Not enough sharp shots

When shooting portraits, the focus point should always be on the eyes. The eyes need to be the sharpest part of your image. In order to achieve sharpness and reduce shake, set your shutter speed to at least one over your focal length. In darker emplacements or even sometimes during the day, it could be helpful to raise your

ISO. A higher ISO will allow you to use a faster shutter speed and a smaller aperture. This way, your entire image will be in focus.

3. Not well composed photographs

For some people it can be rather difficult to get their images straight. To help you avoid that, when you look through the viewfinder, look for the frame or line of reference to straighten your image. It could be simply anything, like a horizon or the pillar. Pay attention if you are lowering your camera on one side. Often people tend to lower it consistently on one side. Try to make a conscious effort to level it.

Check on the horizontals and verticals to keep them level in your viewfinder. You can use the layout grid in the viewfinder to help you do this. If the sensor plane (the back of the camera) is not vertical, you can get the effect called keystoning. Keystoning means the converging of lines that are supposed to be parallel. Sometimes this can be corrected only by using post-processing software or an expensive tilt-shift lens.

Also check on the bits of branches or little pieces of something on the edges of the frame. These distractions cause the viewer's eye to wander away from the main subject. Before taking a shot, adjust a bit your composition. A simple step to one side, or a slight change in camera angle can easily fix the problem.

Pay attention to the intersection of two different visual elements. They usually draw the attention, while this overlapping and touching can be simply avoided by slight changes in your perspective.

Also sometimes it can help to try to simplify your composition so that nothing competes with the subject. This can also be improved by changing your perspective or removing the distracting objects. On other occasions, compositions can be too simple. In this case, create more complex composition by adding more elements.

Instead of wasting hours and hours on post-processing, often you could simply stop for a moment when you see something interesting and think about the best way to capture it before taking a shot.

4. Too far away

Often photographs are not good enough because you were too far away. So come closer and capture your subject large in the frame using a wide angle lens.

5. Problems with contrast, exposure, black and white levels

These are the vital elements of your image, so try to do them well. It is true that you can fix them later and that often you can do it well. But it is much better when done right in the camera.

6. No subject

Besides having a form, your image also needs a content. Beauty, light and color will be emphasized by an interesting subject matter that will give them some substance. Think about subject, idea or emotion that appeals to you, and work on it.

7. Problems with gear

Maybe you need to spend a bit more time to get to know your tools properly to be able to focus on exposure, composition or framing. First fix your technical problems and then focus on making images.

Conclusion

I hope you have enjoyed this book and that it has been a real help for you in improving your photography skills. Thank you for reading it!

A photographer is expressing the world anew with the help of light. To become a photographer, you should therefore first of all learn to see. And to listen. The subject will guide you in your decisions. But you should also learn to listen to your seniors in photography. They have learned the art of photography and they have transmitted to us their experiences. This book has summarized their main rules and best practices.

In the first chapter we have discussed about your camera in order to help you take control of it. We have started with the exposure issues, namely with the exposure triangle: aperture, shutter speed and ISO. These settings are all related to the light, to the way the light interacts with the camera. To learn to set them properly is the basic thing for every image you take. The second topic was the shooting modes. Some of them fully automate the camera exposure and some do it partially, while the manual mode allows you to set them all by yourself. The third topic was the types of lenses. There are five essential types, each of them designed for a specific use. At the end of the first chapter we have explained the other basics of the photography like white balance, metering, focusing, and focal length.

The second chapter was all about the best practices for taking sharp photos. Blurriness is one of the most common mistakes. You should learn how to avoid it and make your pictures look clear and defined. We discussed mainly the issues of slow shutter speed, high ISO number, poor focusing, moving subject, bad lens, and improper hand-holding techniques, but also many others.

In the third chapter we talked about composition. Maybe you have found it to be less technical and more artistic topic. Since the composition directly influences the quality of your images, you should pay particular attention to learn all that you can about the rule of thirds, visual weight, lines, patterns, texture, colors, symmetry, and so on.

The fourth chapter was about improving your portrait, landscape and nature photography. You have probably already had numerous difficulties and issues while trying to capture the most impressive moments. That's why we have laid for you the best tips and practices to help you answer your questions and overcome those difficulties.

In the fifth and the last chapter we have explained the mistakes often made by the DSLR camera users and how to correct them. We talked about the issues of color, sharpness, composition, distance, settings, subjects, and gear.

A famous photographer and environmentalist Ansel Adams once said: "There are no rules for good photographs, there are only good photographs." The rules and the practices explained in this book are only the basics of the photography. The rest is up to you. If you put yourself to work, you will soon be able to discern between a photograph and a good photograph. And you will know what to do about it.

FREE BONUS: "Click The Link Below To Receive Your Bonus

https://publishfs.leadpages.co/pangea-health/

www.ingramcontent.com/pod-product-compliance
Lightning Source LLC
Chambersburg PA
CBHW070136210526
45170CB00013B/1183